A PACK OF DOGS

by Stephanie Fitzgerald

Children's Press®
An imprint of Scholastic Inc.

A special thank-you to
the team at the Cincinnati Zoo & Botanical Garden
for their expert consultation.

Library of Congress Cataloging-in-Publication Data

Names: Fitzgerald, Stephanie, author.
Title: A pack of dogs / by Stephanie Fitzgerald.
Description: New York, NY : Children's Press, an imprint of Scholastic Inc., [2023] | Series: Learn about animals | Includes index.
Audience: Ages 5–7. | Audience: Grades K–1.
Summary: "Next set in the Learn About series about animal groups"—Provided by publisher.
Identifiers: LCCN 2022025176 (print) | LCCN 2022025177 (ebook) | ISBN 9781338853421 (library binding) | ISBN 9781338853438 (paperback) | ISBN 9781338853445 (ebk)
Subjects: LCSH: Wild dogs—Juvenile literature. | Wild dogs—Behavior—Juvenile literature. | Wild dogs—Nomenclature (Popular)—Juvenile literature. | BISAC: JUVENILE NONFICTION / Animals / General | JUVENILE NONFICTION / Animals / Dogs
Classification: LCC QL737.C22 F57 2023 (print) | LCC QL737.C22 (ebook) | DDC 599.77—dc23/eng/20220606
LC record available at https://lccn.loc.gov/2022025176
LC ebook record available at https://lccn.loc.gov/2022025177

Copyright © 2023 by Scholastic Inc.

All rights reserved. Published by Children's Press, an imprint of Scholastic Inc., *Publishers since 1920.* SCHOLASTIC, CHILDREN'S PRESS, and associated logos are trademarks and/or registered trademarks of Scholastic Inc.

The publisher does not have any control over and does not assume any responsibility for author or third-party websites or their content.

No part of this publication may be reproduced, stored in a retrieval system, or transmitted in any form or by any means, electronic, mechanical, photocopying, recording, or otherwise, without written permission of the publisher. For information regarding permission, write to Scholastic Inc., Attention: Permissions Department, 557 Broadway, New York, NY 10012.

10 9 8 7 6 5 4 3 2 1   23 24 25 26 27

Printed in China   62
First edition, 2023

Book design by Kimberly Shake

Photos ©: cover: Jim Cumming/Getty Images; 7 bottom: Donald M. Jones/Minden Pictures; 8–9: Hiroya Minakuchi/Minden Pictures; 10 bottom left: Andyworks/Getty Images; 10 bottom right: Vicki Jauron, Babylon and Beyond Photography/Getty Images; 11 bottom: Tier Und Naturfotografie J und C Sohns/Getty Images; 11 center: aleroy4/Getty Images; 12–13: Tom Mangelsen/NPL/Minden Pictures; 14–15 background: Jim McMahon/Mapman ®; 14 top: Sumio Harada/Minden Pictures; 15 top: Dr P. Marazzi/Science Source; 15 bottom left: Vicki Jauron, Babylon and Beyond Photography/Getty Images; 15 bottom right: Juergen & Christine Sohns/Minden Pictures; 16–17: Dennis Fast/VWPics/Universal Images Group/Getty Images; 20–21: 90706249Wim van den Heever/NPL/Minden Pictures; 22–23: Lynn M. Stone/NPL/Minden Pictures; 24–25: AfriPics.com/Alamy Images; 26 bottom: Hira Punjabi/Alamy Images; 27 top: ozflash/Getty Images; 27 center left: Dave Watts/NPL/Minden Pictures; 27 center right: Roland Seitre/Minden Pictures; 27 bottom: Jurgen and Christine Sohns/FLPA/Minden Pictures; 28 top: Ignacio Yufera/Biosphoto/Minden Pictures; 29 bottom: Jaymi Heimbuch/Minden Pictures; 30 bottom: Larry Williams/Getty Images.

All other photos © Shutterstock.

TABLE OF CONTENTS

INTRODUCTION
What Is a Pack of Dogs? . . . 4

CHAPTER 1
Do All Dogs Live in Packs? . . . 6

CHAPTER 2
Where Do Dog Packs Live? . . . 12

CHAPTER 3
Dog Bodies . . . 16

CHAPTER 4
What Do Wild Dogs Eat? . . . 20

CHAPTER 5
A Pack of Puppies . . . 24

CONCLUSION
Family Ties . . . 28

Wild Dogs at Risk . . . 29

Pet Dogs v. Wolves . . . 30

Glossary . . . 31

Index/About the Author . . . 32

INTRODUCTION

WHAT IS A PACK OF DOGS?

Many animals form groups for different reasons. Some animals travel together in groups. Some form groups to protect one another. Groups of different animals also have many different names. A group of wild dogs is called a **pack**. Wild dogs are members of the dog family called **canines**. They are **predators**. This means they hunt other animals for food.

African wild dogs hunt in the morning and early evening.

There are 34 **species**, or types, of canines.

CHAPTER 1

DO ALL DOGS LIVE IN PACKS?

African wild dogs, dholes, dingoes, and gray wolves are all types of canines that live in packs. The pack works together to take down large **prey**. Other wild dogs, like jackals, eat smaller prey. They do not usually live in packs. Pet dogs do not live in packs, either! They probably consider their family their pack.

Black-backed jackals live in East Africa and southern Africa.

Coyotes live in packs. But they often hunt alone or in pairs.

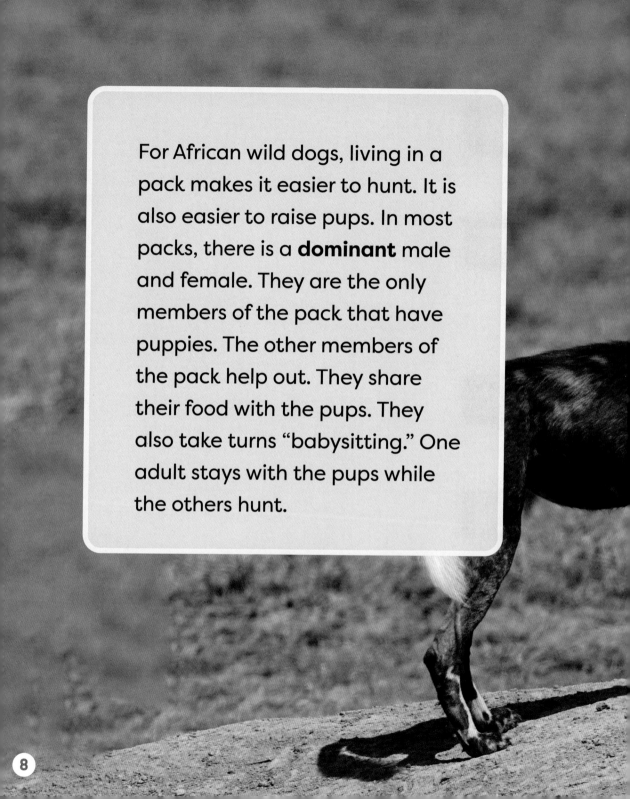

For African wild dogs, living in a pack makes it easier to hunt. It is also easier to raise pups. In most packs, there is a **dominant** male and female. They are the only members of the pack that have puppies. The other members of the pack help out. They share their food with the pups. They also take turns "babysitting." One adult stays with the pups while the others hunt.

A gray wolf pack usually has 4 to 7 members. Each pack has a **unique** howl. That is how the animals communicate. A wolf may howl to tell other members of the pack where it is. African wild dog packs can have up to 20 members. Each animal has a different pattern on its coat. A coyote pack is made up of alphas and their pups. It may also include other adult coyotes.

Gray Wolves

African Wild Dogs

Dholes

Coyotes

Dingoes

A dhole pack has 5 to 12 members. Sometimes several packs join up to hunt. After sharing their prey, the packs go their separate ways. A dingo pack usually has about 10 family members. Dingoes use a variety of howls to communicate.

CHAPTER 2

WHERE DO DOG PACKS LIVE?

Most wild dogs spend most of their time outside. However, they do use simple dens to have babies and raise their pups. A pack of African wild dogs might dig a hole in the ground for the pups. Sometimes they take over the den of another animal. Gray wolves may do the same. They may also use a hollow log or an opening in some rocks for a den. Dholes usually have large underground dens with many entrances.

Wild dogs often change dens at least once before the pups are ready to leave for good.

This litter of coyote pups has a rocky den.

Wild Canine Packs

Coyotes

NORTH AMERICA

SOUTH AMERICA

Packs of wild dogs can be found throughout the world. They live in areas that do not have a lot of people.

This map shows where African wild dogs, coyotes, dholes, dingoes, and gray wolves live.

CHAPTER 3

DOG BODIES

Wild dogs come in a variety of sizes. The gray wolf is a type of wolf. It is the largest canine, or member of the dog family. A gray wolf can grow up to 30 inches (76 cm) tall at the shoulder. It can weigh up to 140 pounds (64 kg). That's about the same weight as an adult human!

Wolves and dogs pant to cool down.

Gray wolves can be lots of different colors.

A Gray Wolf's Body

BUSHY TAILS
A gray wolf uses its tail to communicate. The position of the tail shows how the animal is feeling.

FUR
Gray wolves are covered in fur.

Only the African wild dog has four toes in front and back.

FEET
Most gray wolves have four toes on their back feet and five toes in front. They walk and run on their toes.

EARS
Gray wolves have very good hearing. They can move their ears back and forth. That lets them zero in on where a sound is coming from.

MUZZLE
The muzzle is the front part of the face. A gray wolf has a short, wide muzzle.

TEETH
Their teeth are made for tearing and biting.

NOSE
Gray wolves have very sensitive noses. In fact, their sense of smell is 100 times stronger than a human's.

CHAPTER 4

WHAT DO WILD DOGS EAT?

African wild dogs, dholes, dingoes, and gray wolves are **carnivores**. They eat only meat. Coyotes and jackals are **omnivores**. They eat plants and meat. What each species eats depends on where it lives. Dog packs mostly go after large prey. Many of them eat deer or antelopes. Wild dogs that hunt alone go after smaller prey, like rabbits and mice.

Impala are African wild dogs' favorite prey.

The African wild dog is the most successful hunter. It can usually catch its prey in 8 tries out of 10.

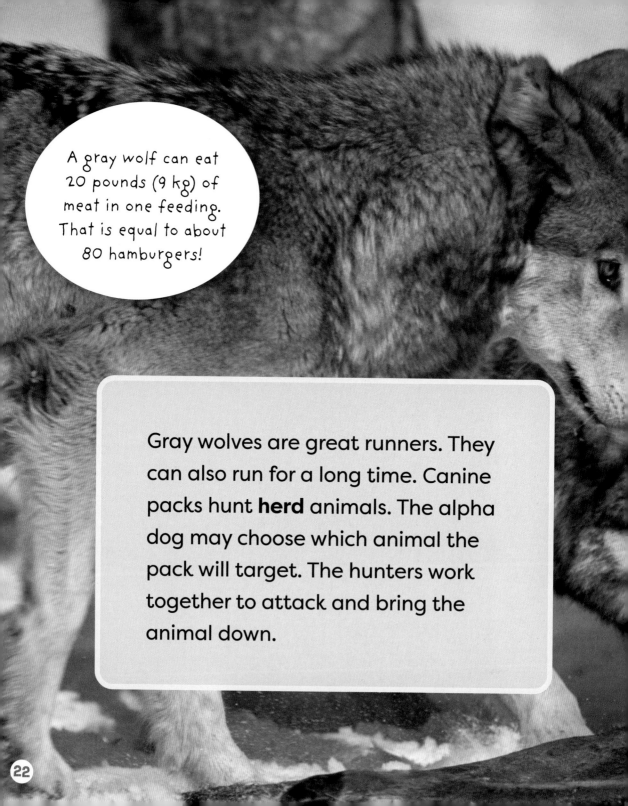

A gray wolf can eat 20 pounds (9 kg) of meat in one feeding. That is equal to about 80 hamburgers!

Gray wolves are great runners. They can also run for a long time. Canine packs hunt **herd** animals. The alpha dog may choose which animal the pack will target. The hunters work together to attack and bring the animal down.

A wolf pack shares a meal.

CHAPTER 5

A PACK OF PUPPIES

Only one pair of wild dogs in a pack will have babies, called pups. Most species have 4 to 7 pups in a **litter**. African wild dogs have about 10 pups in a litter. Dholes can have up to 12. At first, the pups drink their mother's milk. After about a month, they can eat meat. The adults bring meat back from a hunt. They chew, swallow, and spit the meat back up to make it easier for the pups to eat.

Pups drink their mom's milk.

Pups lick the adults' faces and jaws to be fed.

Pups are born helpless and blind. They are completely dependent on the pack for their survival. But, wild dogs grow up fast! In just a few months, the pups are out of the den. They explore and play with their siblings.

Dholes make a variety of sounds, including whistles and screams. They even cluck.

Life Cycle of a Dingo

PUP
Dingo pups will stay in their den for up to three months.

JUVENILE
Within their first year, dingoes are ready to hunt.

ADULT
Dingoes become adults around their first birthday and can have pups of their own.

A dhole greets its pack mates.

CONCLUSION

FAMILY TIES

Now you know more about wild dogs and packs. A pack of dogs is a family. The members work together. They play together. They even take care of one another—from sharing food to "babysitting" pups. Wild dog packs have a strong family bond, just like humans!

The dog was the first animal to be **domesticated.**

Wild Dogs at Risk

Wild dogs face the same threats as many other wild animals. Some wild dogs are **endangered**. A dog pack usually needs a large area for hunting. Humans have taken over a lot of land. They build homes, farms, and roads. The open spaces for wild dogs have become smaller and smaller. Farmers may see these wild dogs as a threat to their livestock. Sometimes wild dogs are killed by farmers who want to protect their own animals. How can you help? You can talk to your family about what you learned. Together, you can support organizations that work to protect wild animals.

Coyotes have learned to survive in cities.

Pet Dogs v. Wolves

Pet dogs do share some **traits** with wolves. Like wolves, pet dogs are very social. But pets form strong bonds with people. Wolves form strong bonds with their pack mates. The biggest difference between dogs and wolves is that dogs are domesticated. They live with and rely on humans for their survival. Pets also have different bodies from wolves. For one thing, they are smaller. And most pet dogs have softer fur, floppier ears, and bigger heads than wolves.

Wolves eat only meat. Pet dogs are not as picky!

Pet dogs love their humans!

Glossary

alpha (AL-fuh) socially dominant, especially in a group of animals

canine (KAY-nine) of or having to do with dogs

carnivore (KAHR-nuh-vor) an animal that eats meat

domesticate (duh-MES-ti-kate) to raise an animal so it can live with or be used by people

dominant (DAH-muh-nuhnt) most influential or powerful

endangered (en-DAYN-jurd) an animal or plant that is at risk of becoming extinct

herd (hurd) a large number of animals that stay together or move together

litter (LIT-ur) a number of baby animals that are born at the same time to the same mother

omnivore (AHM-nuh-vor) an animal that eats both plants and meat

pack (pak) a group of often predatory animals of the same kind

predator (PRED-uh-tur) an animal that lives by hunting other animals for food

prey (pray) an animal that is hunted by another animal for food

species (SPEE-sheez) a group of related animals or plants

trait (trayt) a quality or characteristic that makes one person, animal, or thing different from another

unique (yoo-NEEK) being the only one of its kind; unlike anything else

Index

Alpha animals, 9–10, 22

Bodies, 16–19, 30

Canines, 4–6, 16, 22
carnivores, 20

Dens, 12–13, 26–27
diets, 20–25, 30
domesticated animals, 28, 30
dominant animals, 8–9

Endangered animals, 29

Family bonds, 6, 28, 30
fur, 10, 17–18

Groups of dogs. *See* packs

Habitats, 12–15
herd animals, 22
howling, 10–11

Juvenile dogs, 27

Life cycle, 24–27
litters, 13, 24

Omnivores, 20

Packs
 compared to living alone, 6–7
 defined, 4
 family bonds in, 6, 28, 30
 number of dogs in, 10–11
 reasons for forming, 8–9
 types of dogs in, 6–7
pet dogs, 6, 30
predators, 4
prey, 6, 20–23

pups, 24–28
 adults babysitting, 8–9, 28
 dens for, 12–13, 26–27
 diet of, 24–25
 life cycle of, 24–27
 number per litter, 24
 parents of, 8–10

Species, 5

Traits, 30

Unique features, 10

ABOUT THE AUTHOR

Stephanie Fitzgerald has been writing nonfiction books for kids for more than 25 years. She thinks the best thing about being an author is that she's always learning something new.